My Skin Is Bumpy and Slimy

by Jessica Rudolph

Consultants:
Christopher Kuhar, PhD
Executive Director
Cleveland Metroparks Zoo
Cleveland, Ohio

Kimberly Brenneman, PhD
National Institute for Early Education Research
Rutgers University
New Brunswick, New Jersey

BEARPORT PUBLISHING

New York, New York

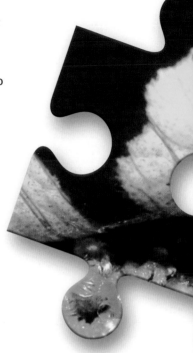

Credits
Cover, © Marcos Veiga/Alamy; 4–5, © Biosphoto/SuperStock; 6–7, © Jason Mintzer;
8–9, © iStockphoto/Thinkstock; 10–11, © NHPA/SuperStock; 12–13, © PhotoAlto/
Odilon Dimier; 14–15, © Maximilian Weinzierl/Alamy; 16–17, © F Rauschenbach/
F1 ONLINE/SuperStock; 18–19, © iStockphoto/Thinkstock; 20–21, © iStockphoto/
Thinkstock; 22, © NaturePL/SuperStock; 23, © iStockphoto/Thinkstock; 24, © Marko
Masterl (MYN)/Nature Picture Library/Corbis.

Publisher: Kenn Goin
Creative Director: Spencer Brinker
Design: Debrah Kaiser
Photo Researcher: We Research Pictures, LLC

Library of Congress Cataloging-in-Publication Data

Rudolph, Jessica, author.
  My skin is bumpy and slimy / by Jessica Rudolph ; consultant, Christopher Kuhar, PhD,
Executive Director Cleveland Metroparks Zoo, Cleveland, Ohio.
       pages cm. — (Zoo clues)
  Audience: Ages 5–8.
  Includes bibliographical references and index.
  ISBN 978-1-62724-115-1 (library binding) — ISBN 1-62724-115-9 (library binding)
  1. Salamanders—Juvenile literature. I. Kuhar, Christopher, consultant. II. Title.
  QL668.C2R83 2014
  597.8'5—dc23
                         2013037790

For more information, write to Bearport Publishing Company, Inc., 45 West 21st Street, Suite 3B,
New York, New York 10010. Printed in the United States of America.

10 9 8 7 6 5 4 3 2

Contents

What Am I?

Look at my eyes.

They stick out
of my head.

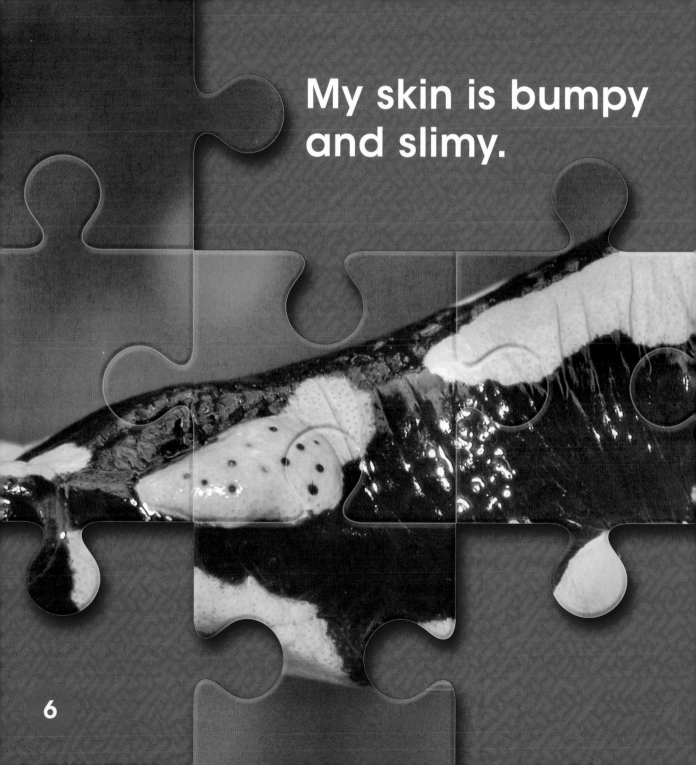

My skin is bumpy and slimy.

6

It is black and yellow.

My tail is long.

8

It is thin at
the end.

9

I have short toes.

11

My belly is low
to the ground.

12

14

I have a wide mouth.

My tongue is pink.

What am I?

Let's find out!

I am a fire
salamander!

Animal Facts

Fire salamanders are amphibians. Like all amphibians, they live part of their lives in water and part on land. Unlike most amphibians, fire salamanders give birth to live young instead of laying eggs.

More Fire Salamander Facts

Food:	Insects, spiders, earthworms, slugs, and young frogs
Size:	About 8 inches (20.3 cm) long, including the tail
Weight:	1.4 ounces (40 grams)
Life Span:	About 14 years in the wild
Cool Fact:	A fire salamander has tiny holes on its back that ooze poison. The poison can harm any animal that tries to eat the salamander.

Adult Fire Salamander Size

Where Do I Live?

Fire salamanders live near ponds in Europe. During the day, they hide under rocks, leaves, and logs to escape the sun's heat.

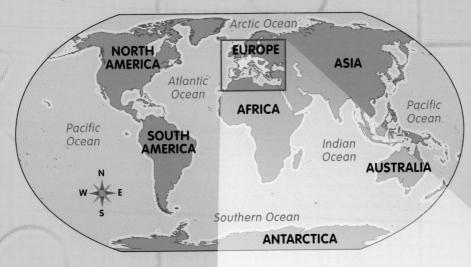

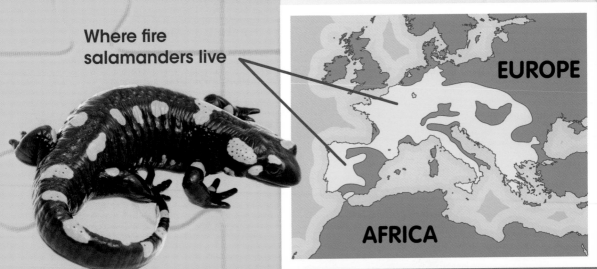

Where fire salamanders live

Index

Read More

Bredeson, Carmen. *Fun Facts About Salamanders.* Berkeley Heights, NJ: Enslow (2008).

Kolpin, Molly. *Salamanders (Pebble Plus: Amphibians).* Mankato, MN: Capstone (2010).

Learn More Online

To learn more about fire salamanders, visit **www.bearportpublishing.com/ZooClues**

About the Author

Jessica Rudolph lives in Connecticut. She has edited and written many books about history, science, and nature for children.